KB245329

휴양하기 좋은
호텔, 리조트 디자인

휴양하기 좋은 호텔,리조트 디자인

발행일 : 2025년 11월 25일
출판사 : 하랑출판
주   소 : 서울시 중구 퇴계로28길 8
전   화 : 02. 2263. 3337

CONTENTS

BOMY
PRESTIGE
港澳
名品 gangaomingpin

This case is a fashion store of high-grade brands with luxurious designings.

storage
fitting room
fitting room
mirror
mirror
counter
furniture
furniture
furniture
display space
entrance
display space

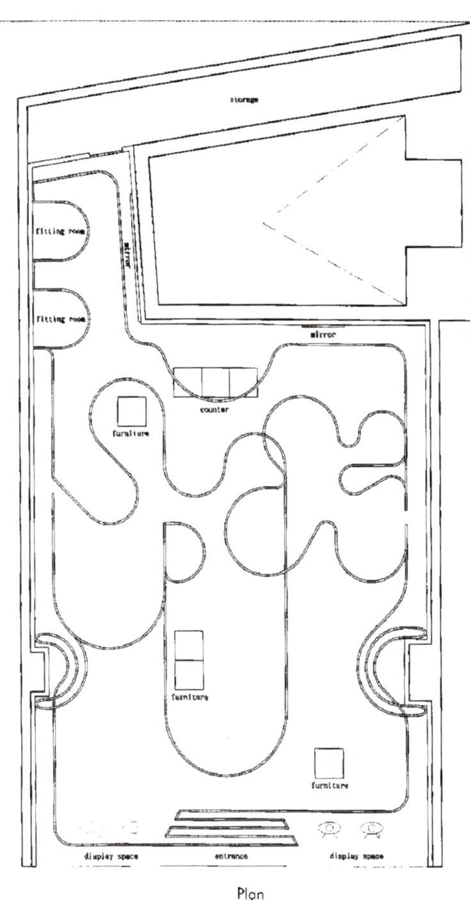

Plan

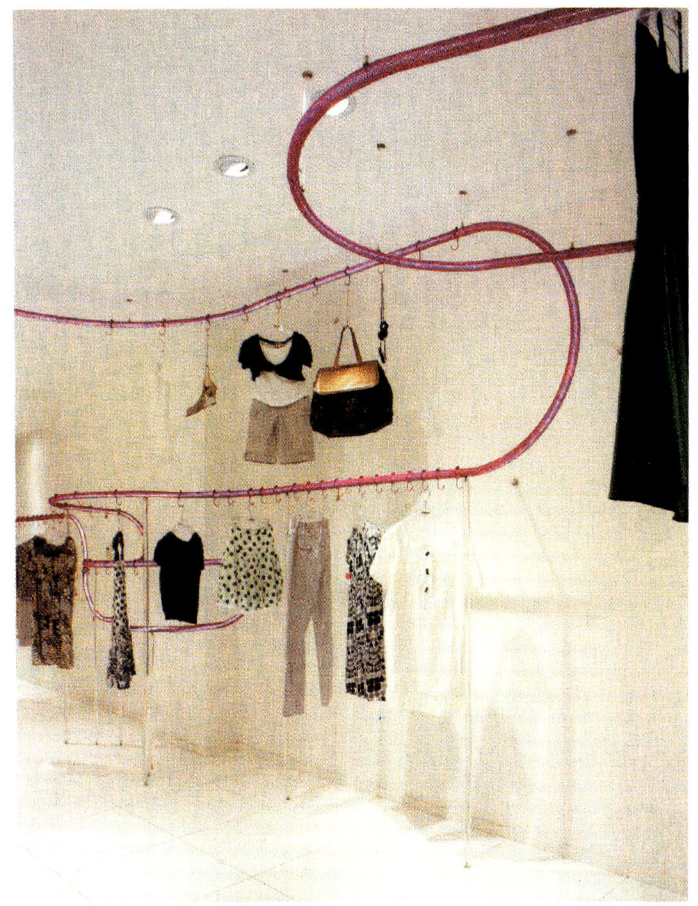

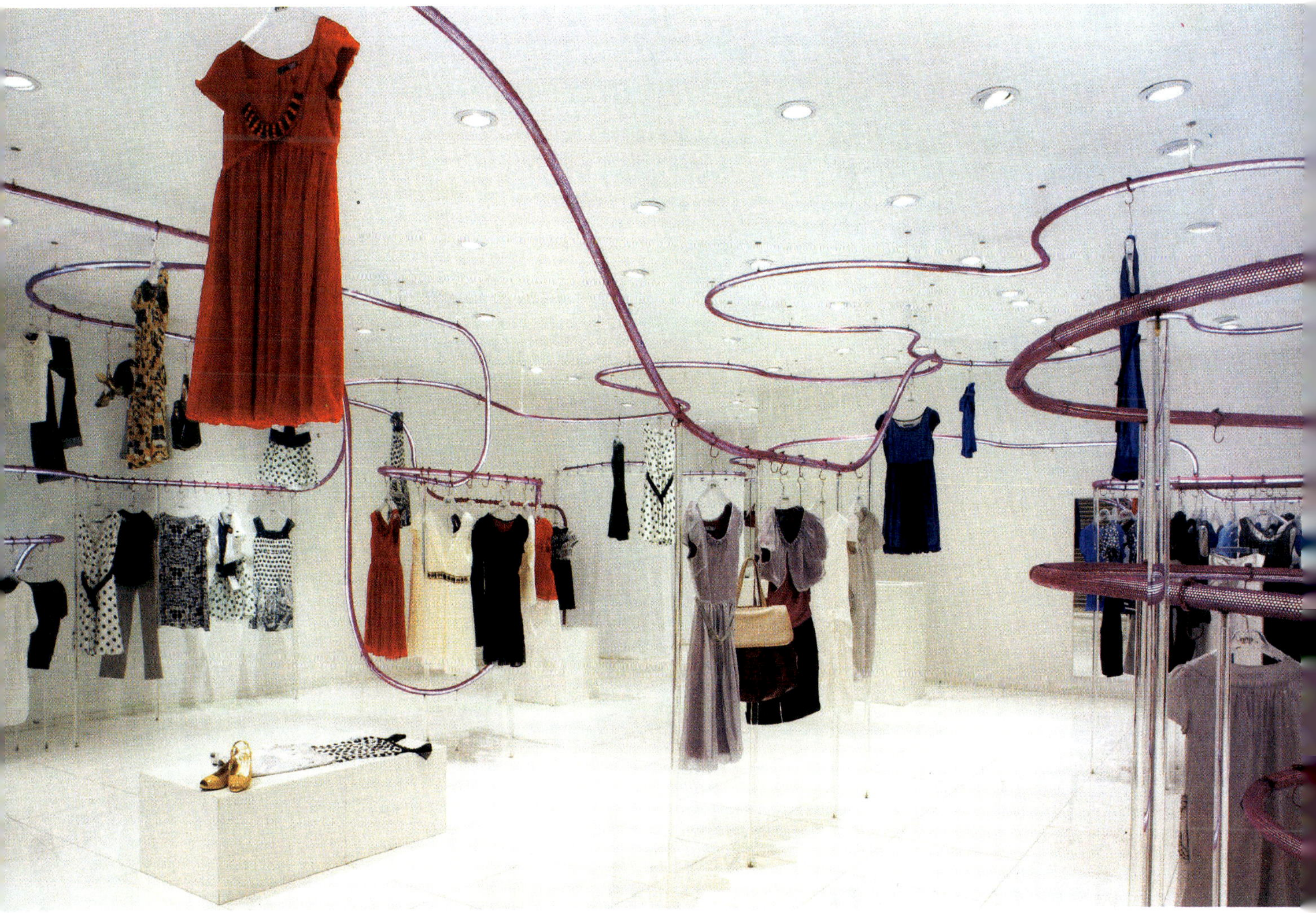

WTCMORE MOMO SHOP

To arouse the shopping interest at one of the major shopping centre in Causeway Bay, Hong Kong, we designed and built a temporary free-standing structure – the MOMO Shop.

MOMO, in Japanese, refers to a kind of fresh juicy peach. The MOMO Shop was created on this concept to attract the youngster/teen-pop fashion market.

To create a mood of "freshly delivered" from Japan, green design is adopted to construct the MOMO Shop. A snatchy structure was incorporated in the design to simultaneously symbolize the creative and unconventional particularity among the young generation.

To achieve a green design, designers selected recycled materials, including recycled wood, MDF board and wrapping bubbles paper, etc. to construct the structure. Moreover, the construction was processed in parts to facilitate the dismantlement and reconstruction afterwards.

Due to the innovative green design and structural characteristics, people attention was attracted while shopping interest was aroused. Following MOMO, three other brands had also adopted the unique design concepts to achieve further advancement.

Irregular and lively architecture shapes made of environment-friendly materials.

MOMO shop made of hundreds of independent parts, random but sensible, fashionable and unique.

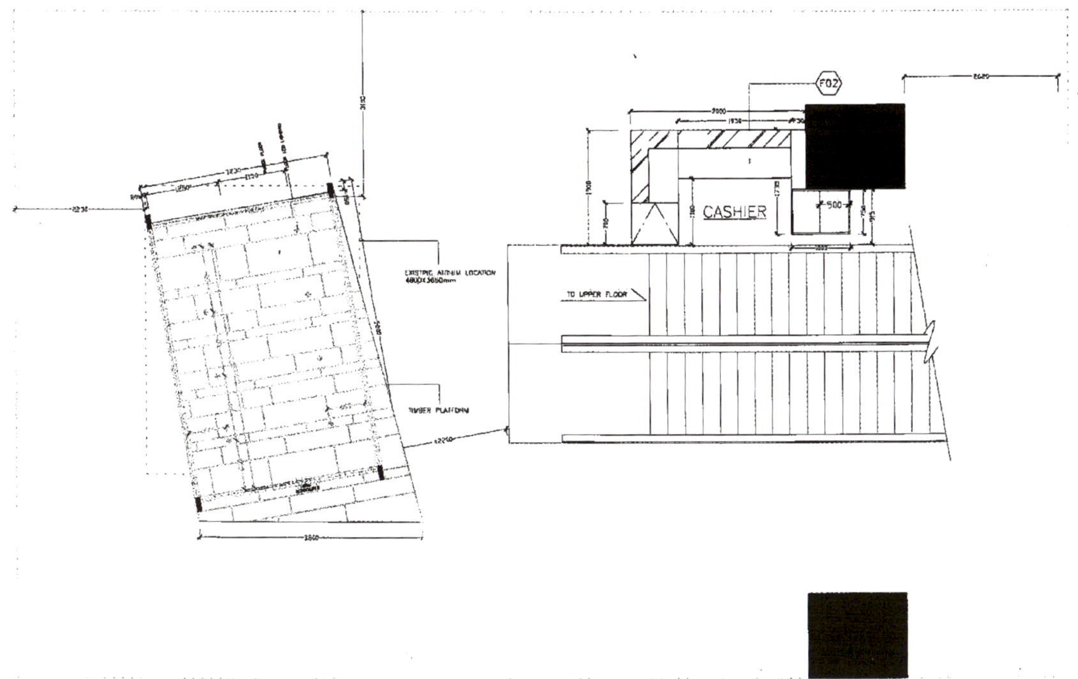

Plan

momo
cashier

EGOIST
SUMMER
SPECIAL
20% Off
Sale
Army of No
Egoist
Avie
Burberry Black Label
is x Mune
Kiit
is x Masaki
Indio
Cicala
Angel City Label
A.T.C
Angel
Burberry Blue Label

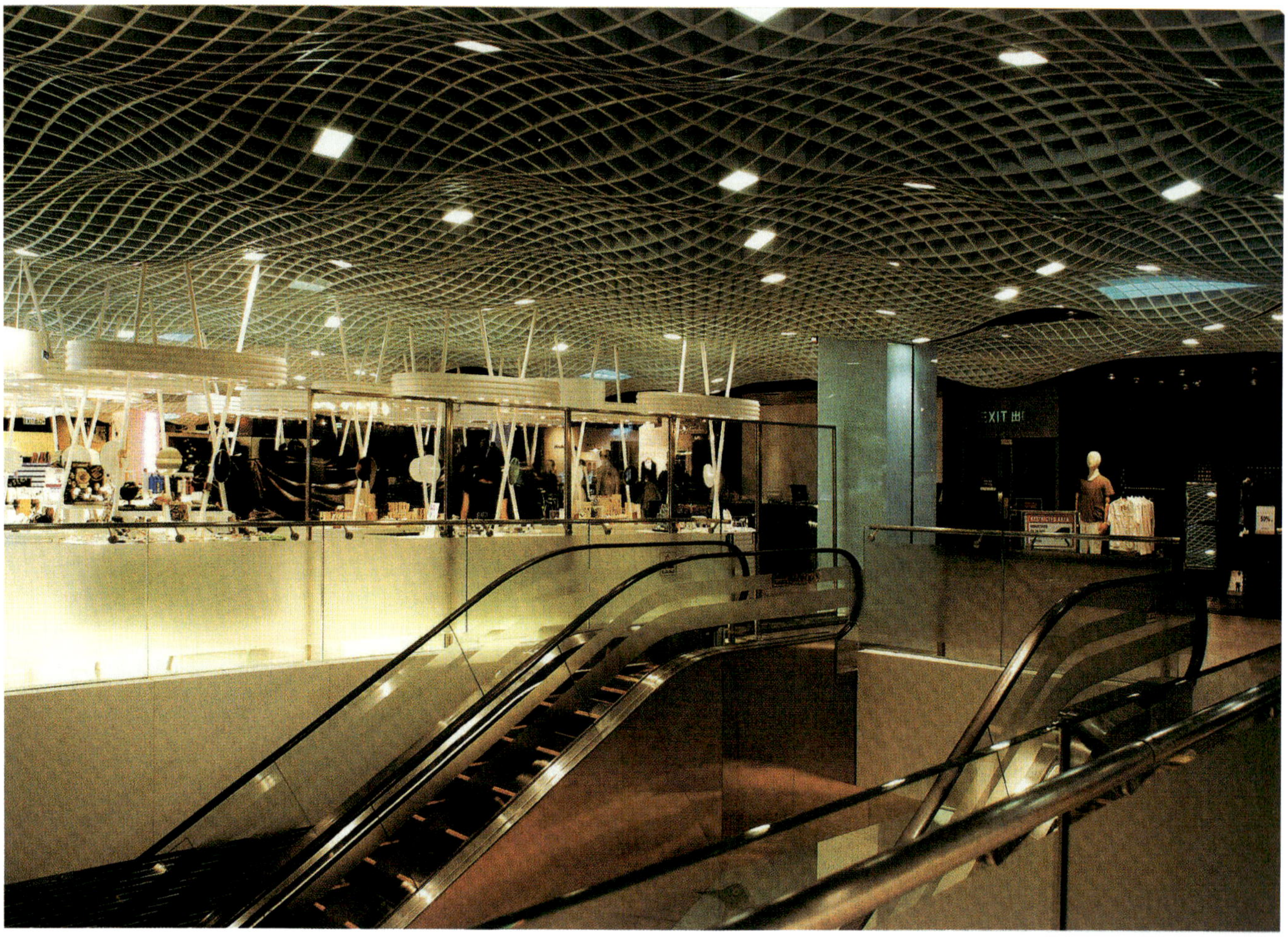

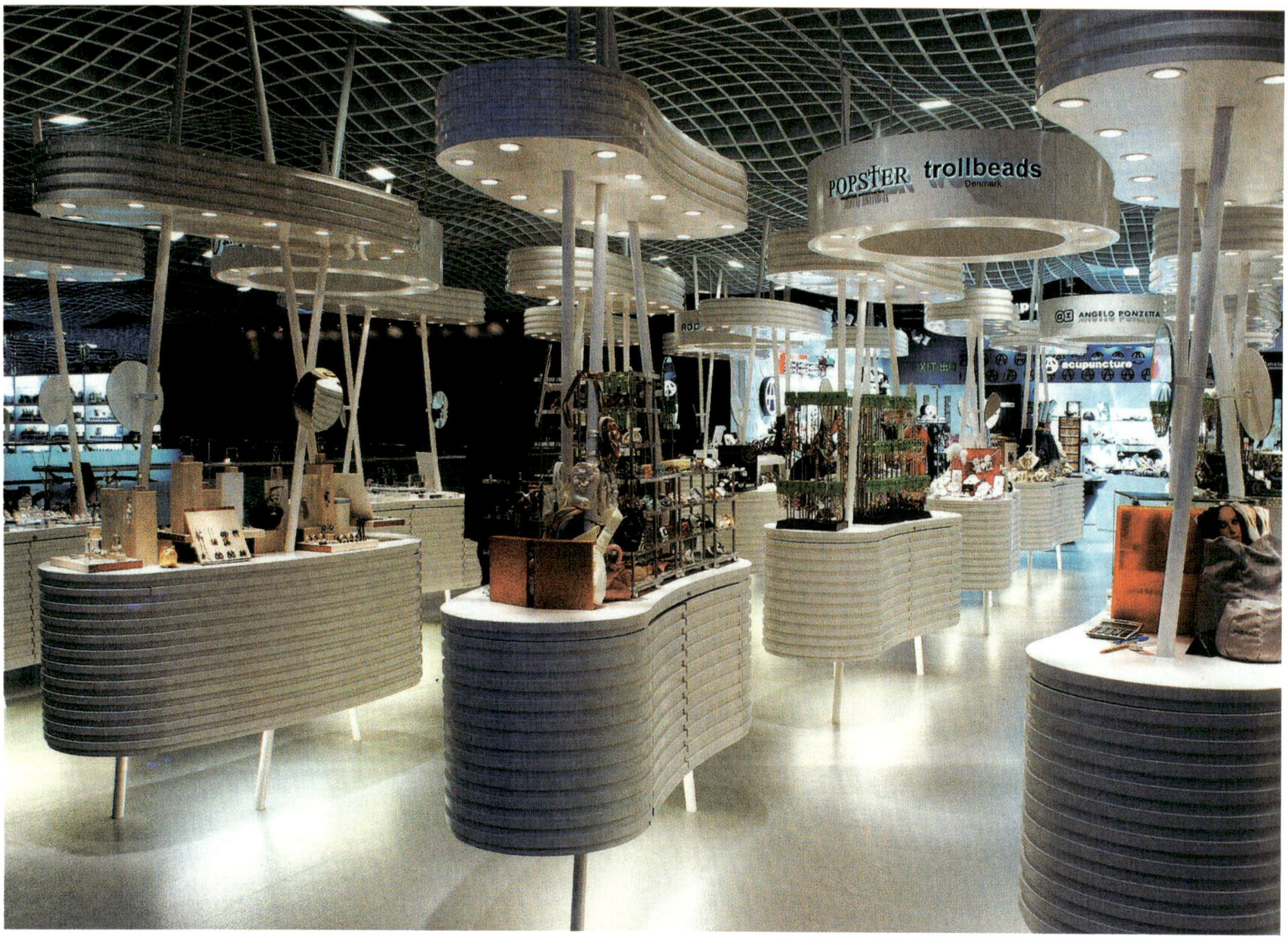
POPSTER
trollbeads
Denmark
ANGELO PONZETTA
acupuncture

Left: The tone of indoors is marine views, with seaside leisure and holiday sense melt in every designing.

Right: Cafe inside the store, a place for relaxation.

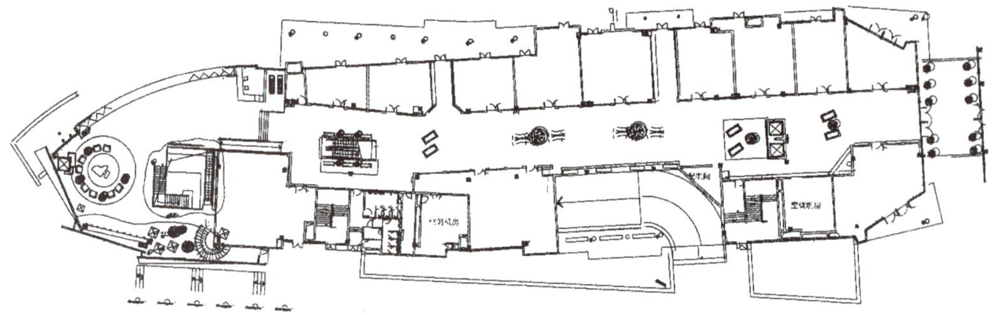

First Floor Plan

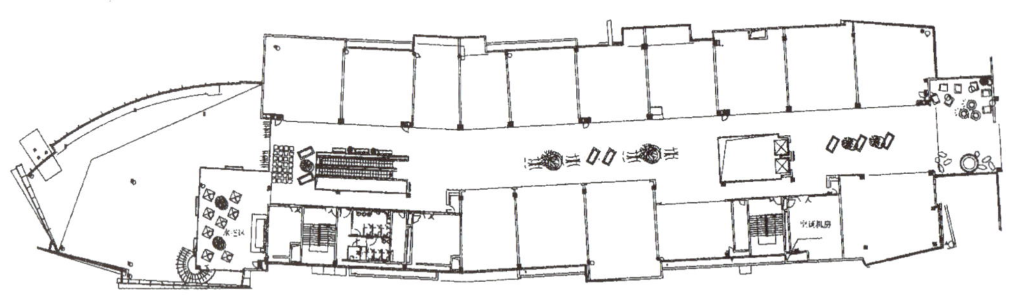

Second Floor Plan

ROEN

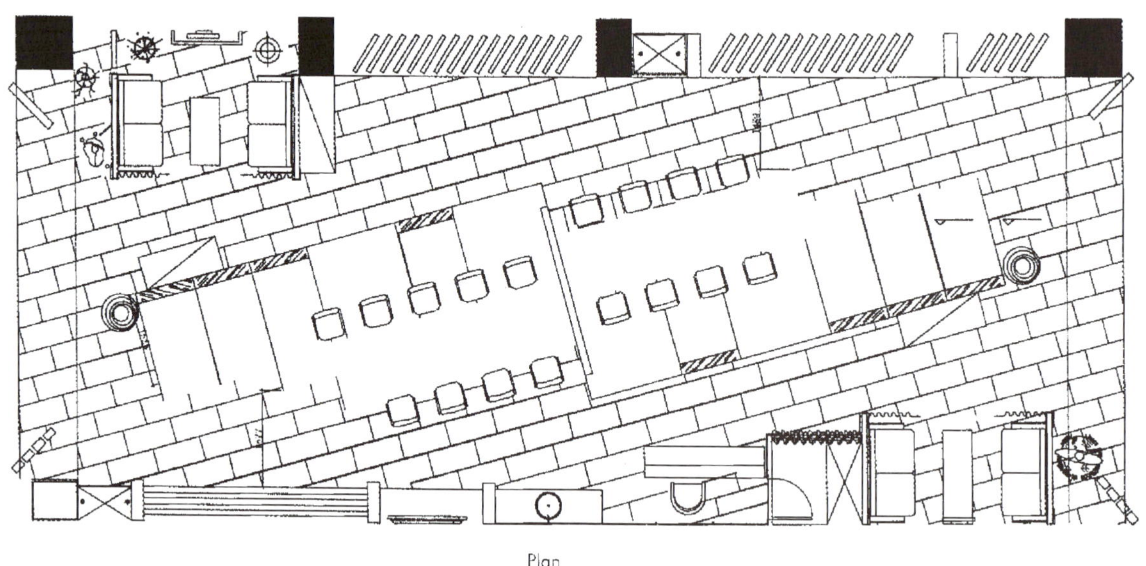

Plan

Oaken walls and molding oak ceilings add a touch to the grand reception hall.

Every piece of the hanging clothe is the
beautiful note forming the melody.

38

A sample store with abundant
merchandises, a perfect place

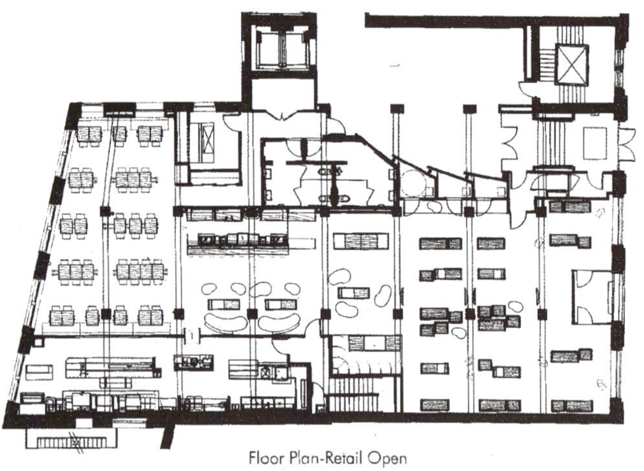

Floor Plan-Retail Open

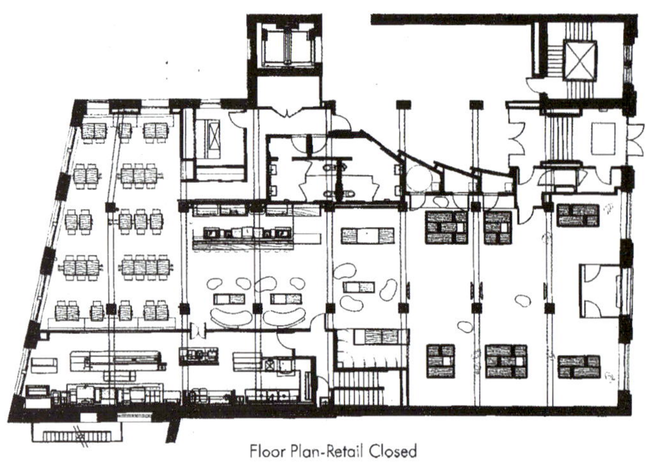

Floor Plan-Retail Closed

SIZES TONIGHT
3 @ 32
ALSO: RED CURRY
ON THE HALF SHELL
ISLAND CREEK
LONG COVE → STONINGTON, ME
CARAQUET → NEW BRUNSWICK
BEAUSOLEIL → NEW BRUNSWICK

Left: The black reception dest with unique patterns represents beauty to each other.
Right: Quiet space for shopping.

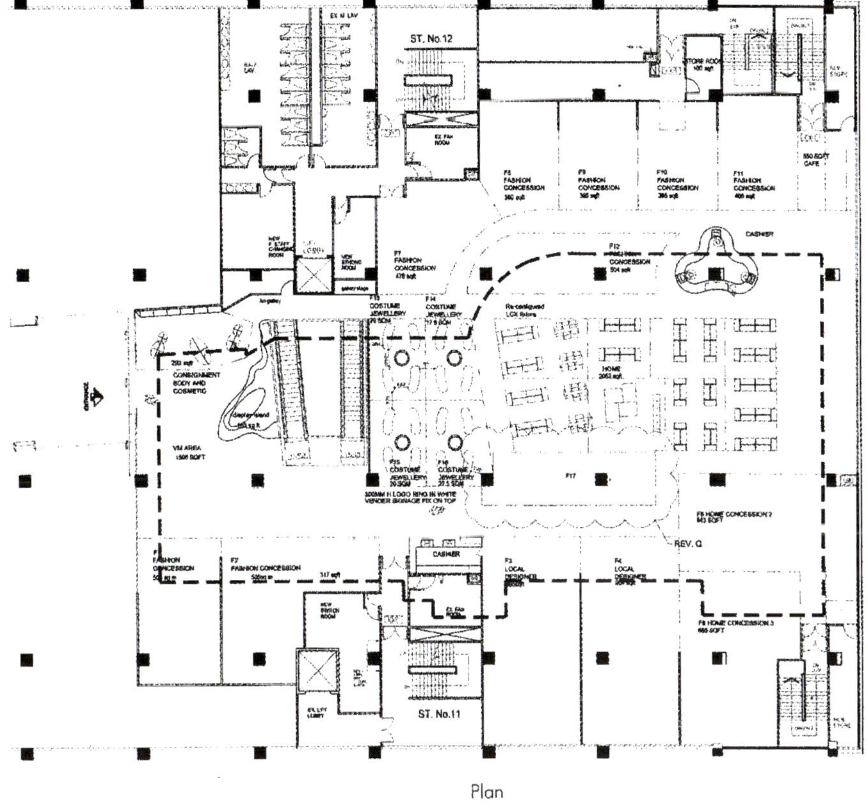

Plan

Blue wavy ceiling models on the waves of the sea while the merchadise on show are the shelis of the sea.

xplus
lavera
ORGANIC
SKIN CARE

EQ:IQ

Mirror
oller shutter
Logo (refer to TD17.1)
1458
1640
LIFT
UP
DN
Typical Timber Rack
7070
Mirror
Mirror
Chang
2057
B
A
Wooden Beach TD-30
500
1800
eat
Existing Circular Light Through
Existing Flas Ceiling Line
D
E
D
A
A
B
C
2074
Feature Display
1400

Plan

EQ:IQ
EQ:IQ

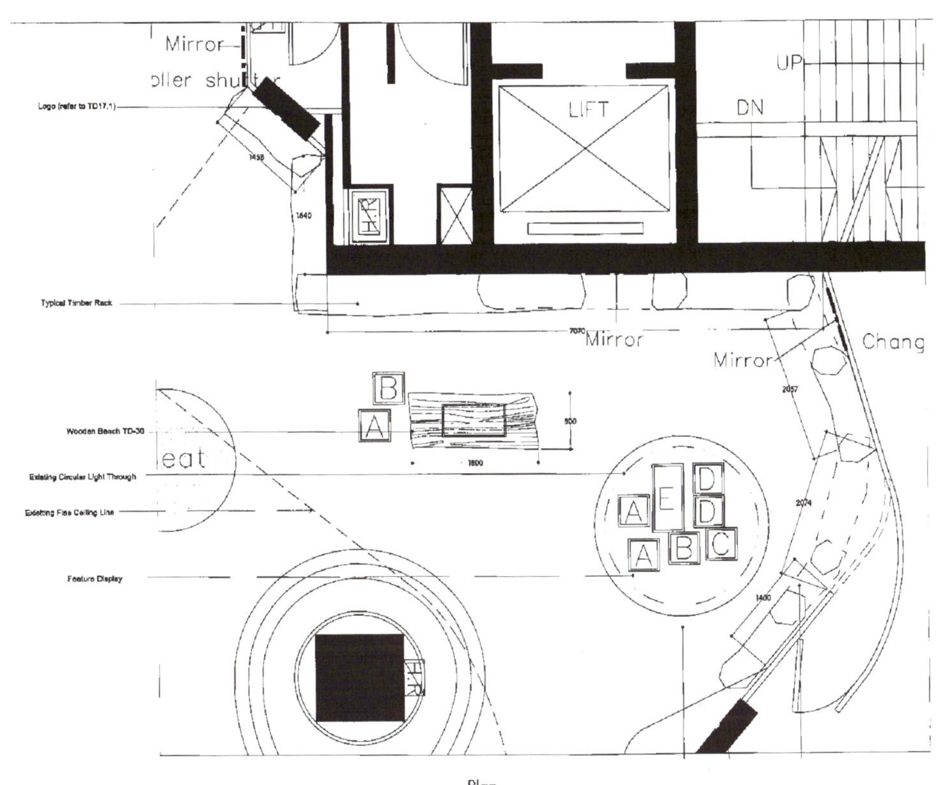

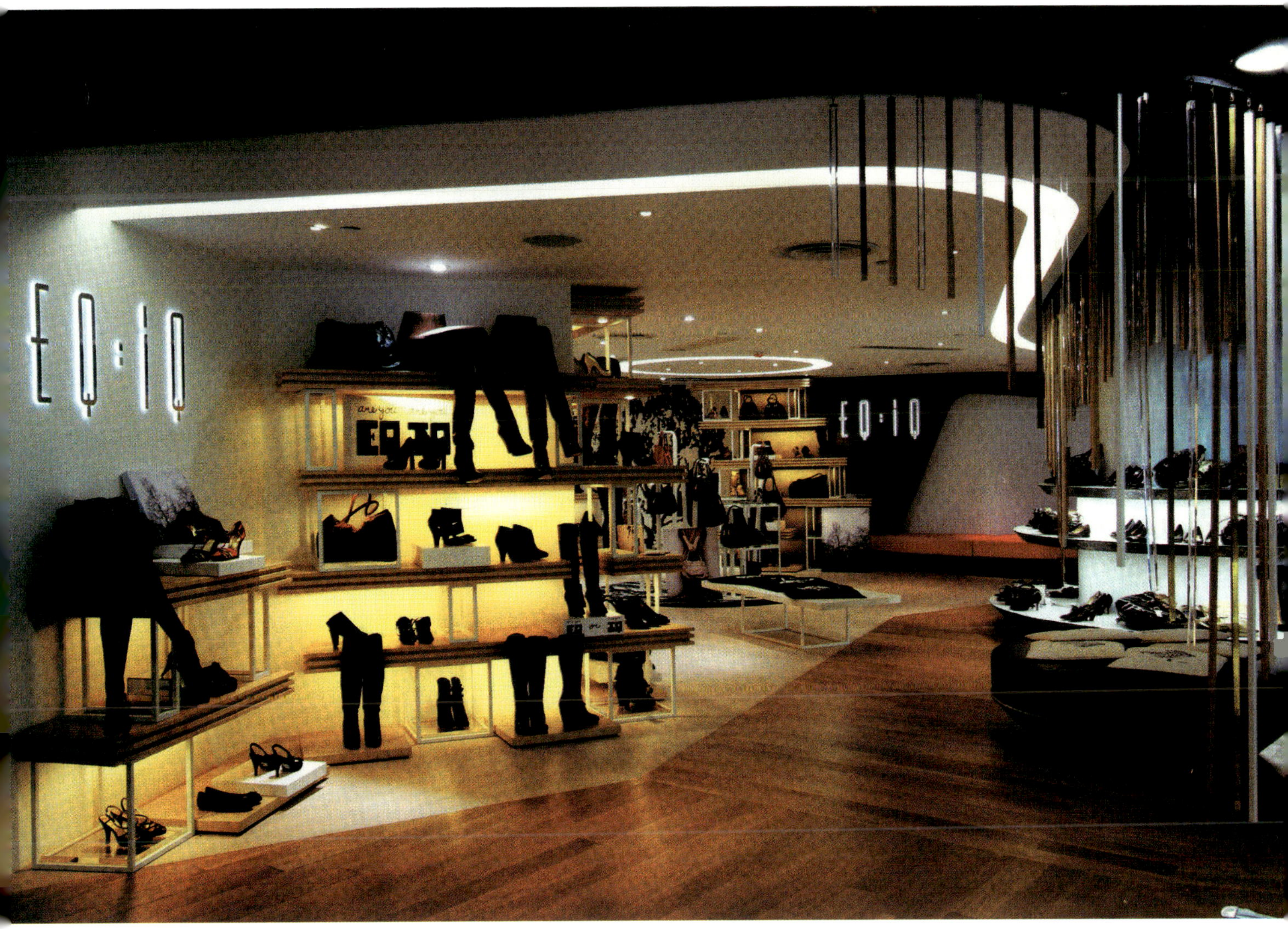

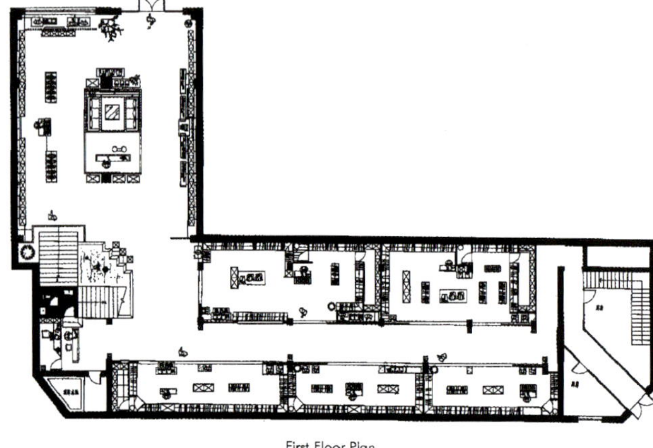

Displays of various brands.

First Floor Plan

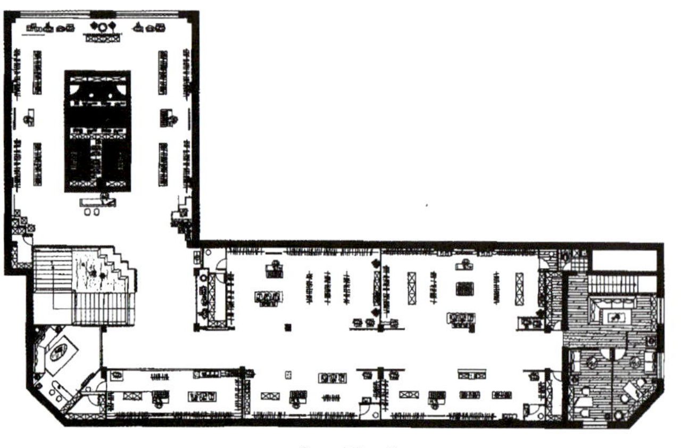
Second Floor Plan

YU HONG SATAM

Delightful arrangement: palette-like overall layout
overall layout and ovate shelters.

Washing room in specially original design is activated by the matching of curving lines and ovate ornaments.

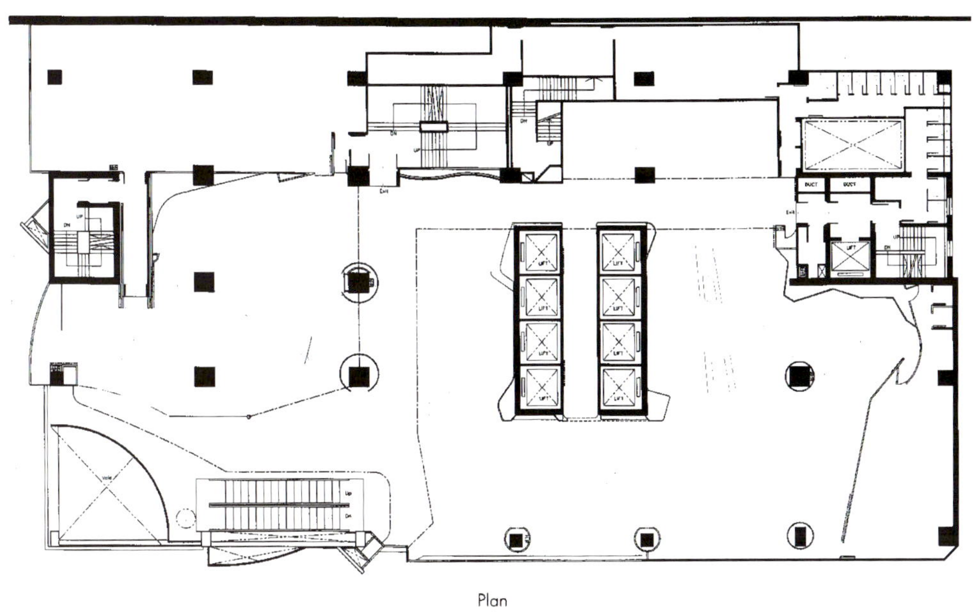

Plan

NINE WEST

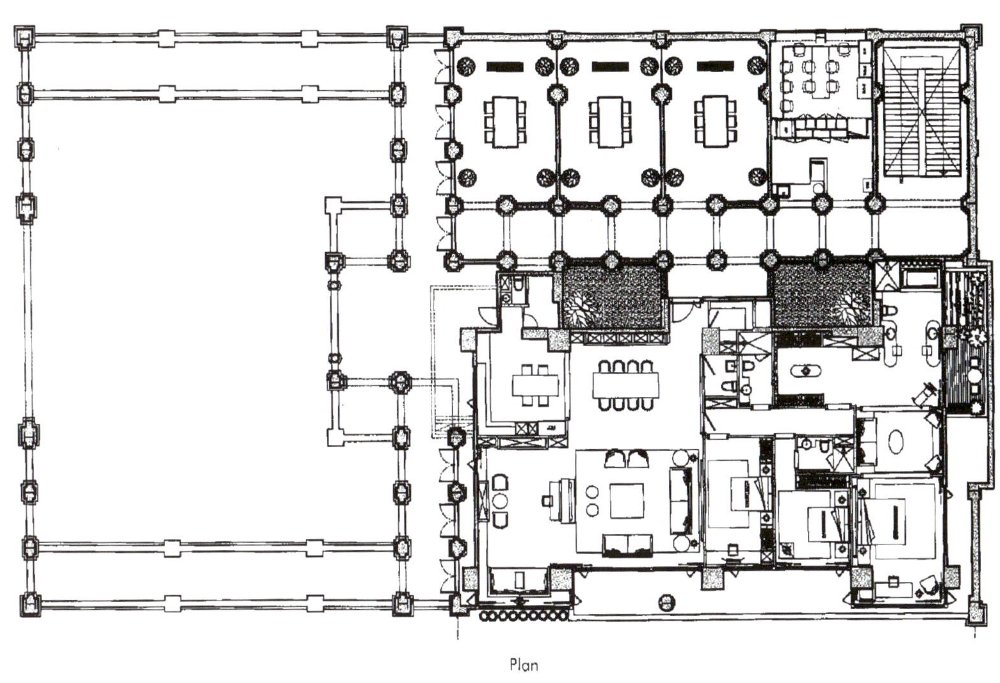

Plan

The stretching arc passageway coupled with towering ceilings are very magnificent.

Left: Terrace sited to command an overall view of the basement.
Right: Luxurious washing room reveals its nobility.

A transparent and bright hall for customers easily recognizing directions.

Left: Abstract lines have a sense of sculpture.
Right: Lights in the room are reflected on the scenic pool.

The passageway with extricate details shows elegance.

Left: Window views beside the stairways.
Right: Discussion area with decorated plants gives out a artistic sense.

富贵世家
Royal Fortune

The graceful stairs and magnificent
Roman columns are all of European style.

富实
Royal

Dainty space decoration with rococo handrails of the staircase and frescos.

Terse molding furniture and artistic fittings
express the quality of life.

The reception center shines magnificently in the light.

The space is filled with the atmosphere of quiet and serene.

Fresh rosy brings a sweetness to the space.

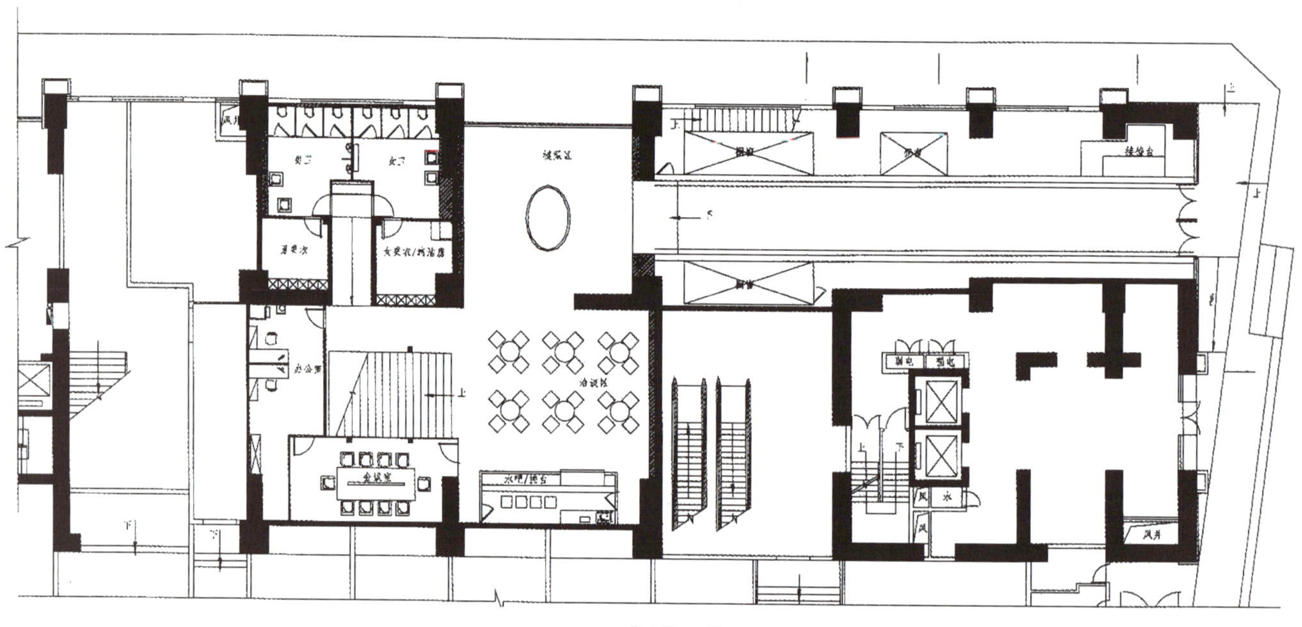

First Floor Plan

Desk-lamp-turned pendant lamps characteriaze the individual barcounter.

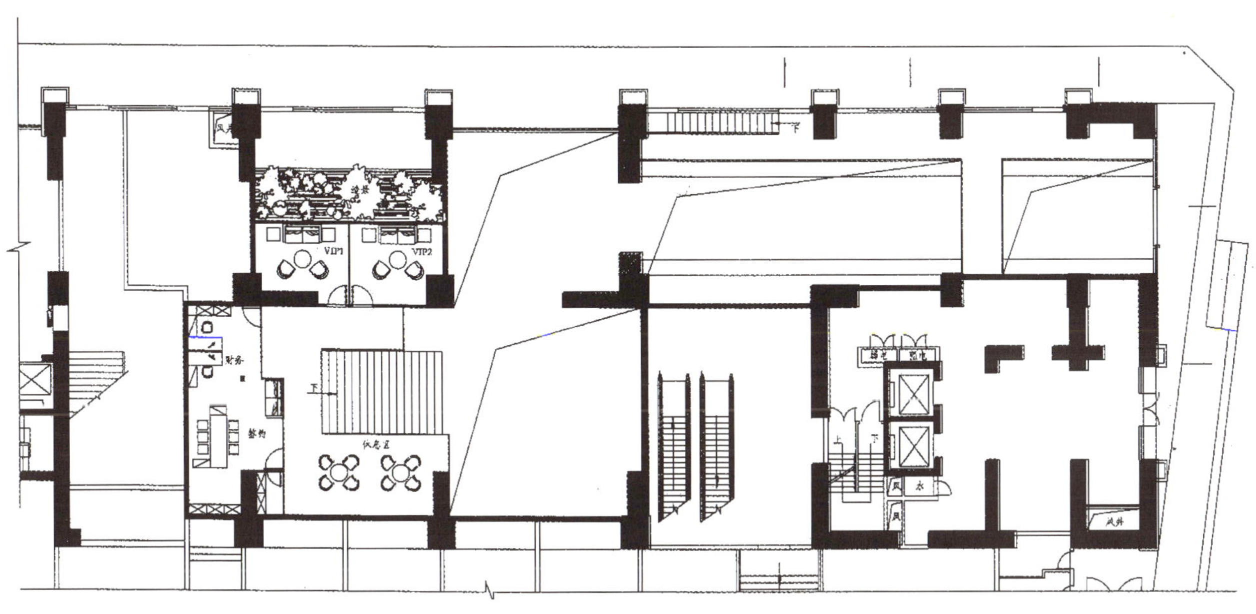

Second Floor Plan

Left: Uneven pendant lamps brighten the area.
Right: Red floods all the washing room except the
white washbasin which will attack your eyes.

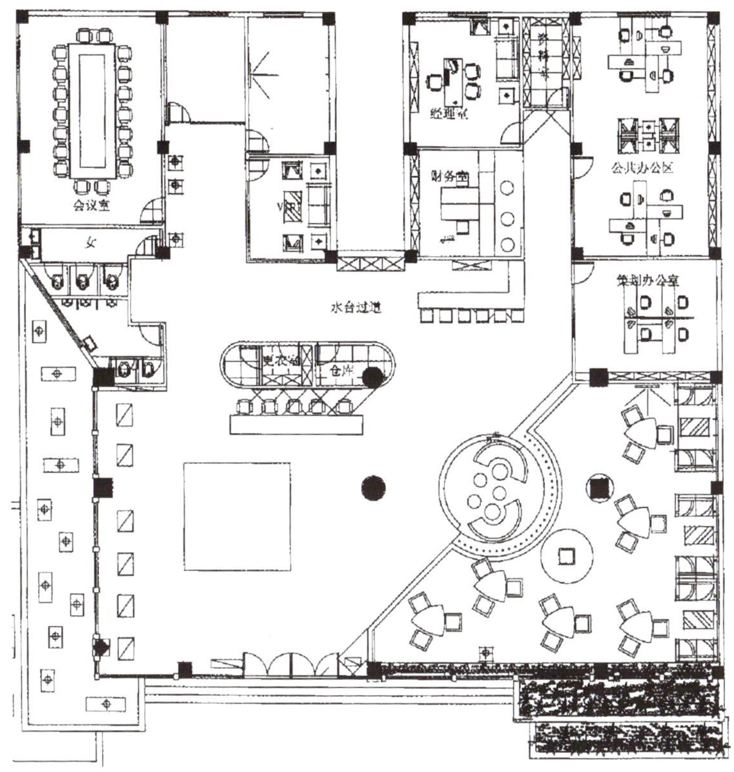

Plan

Steel ball curtains and funny arm chairs make the space simple and charming.

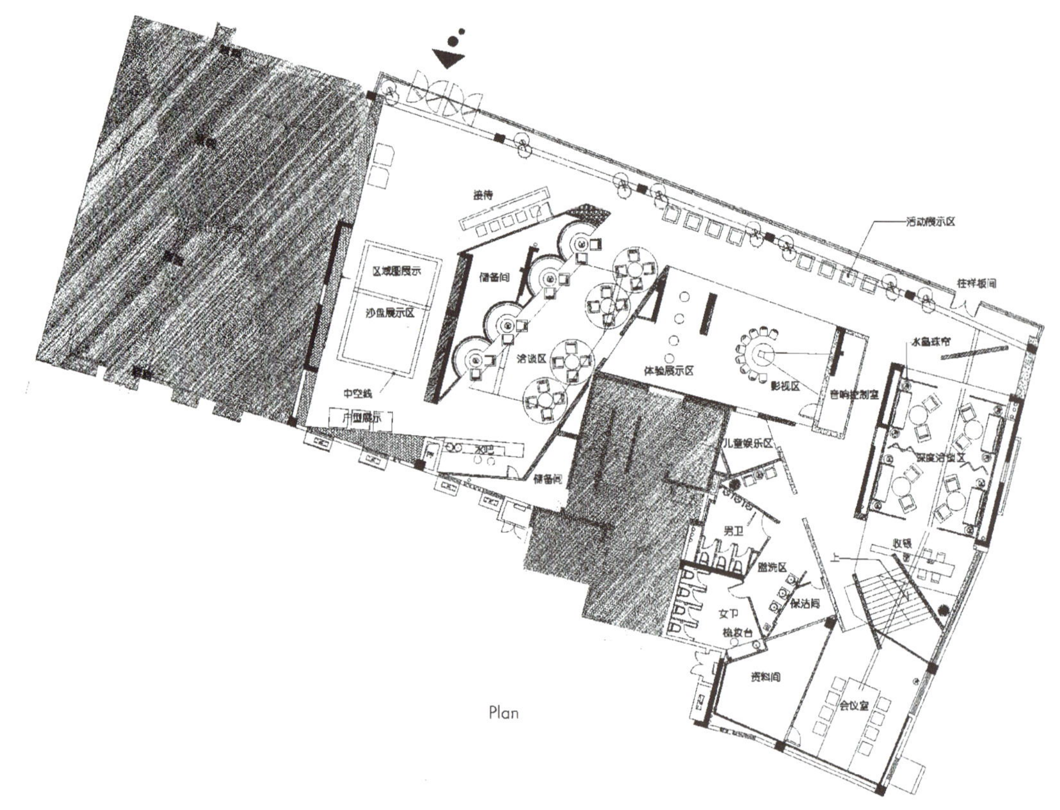

Plan

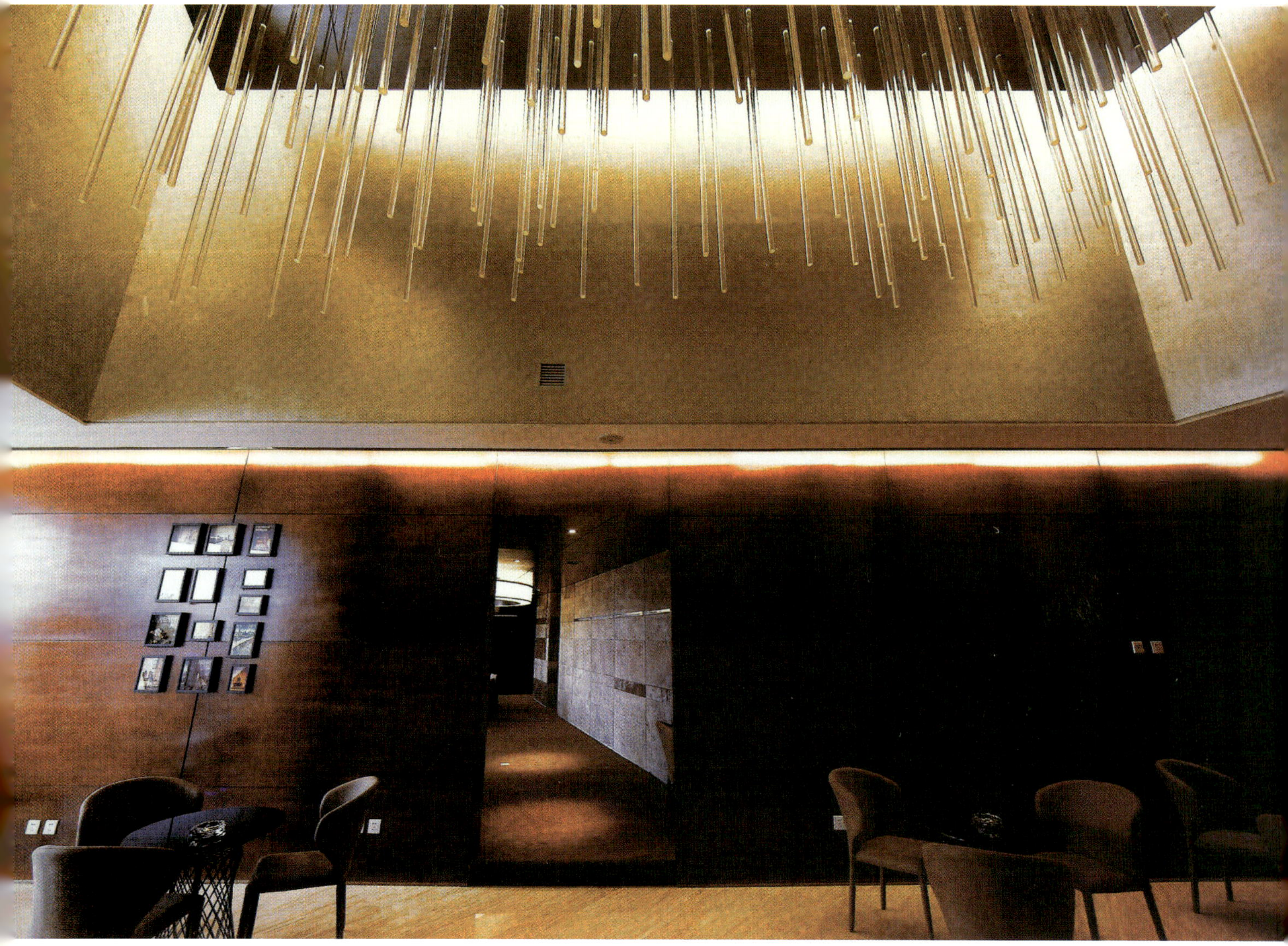

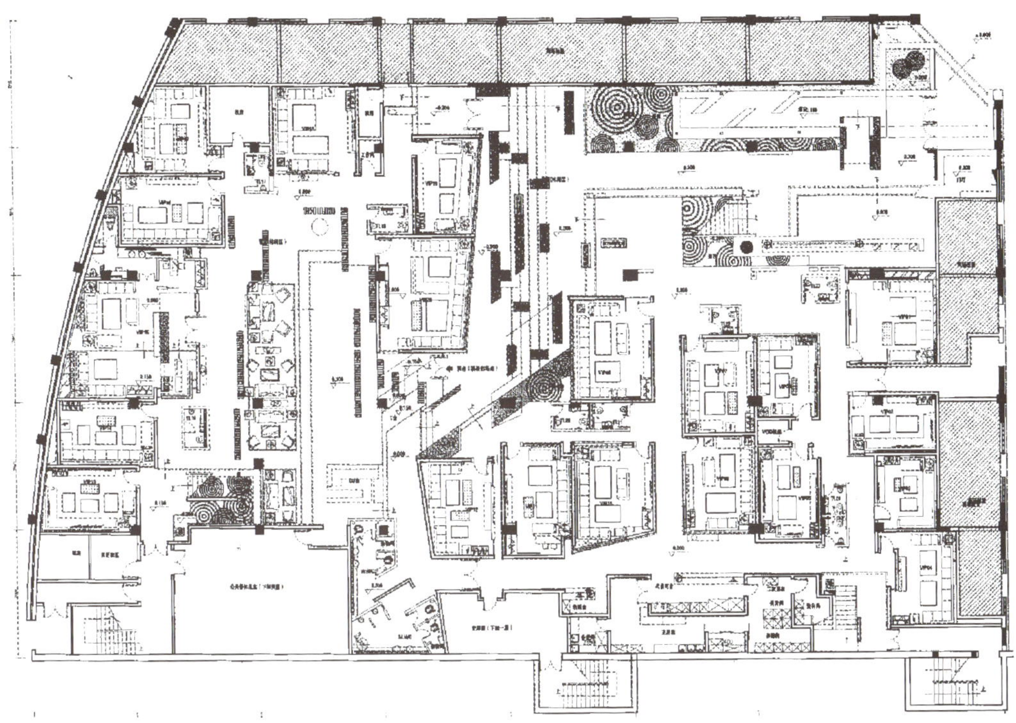

Plan

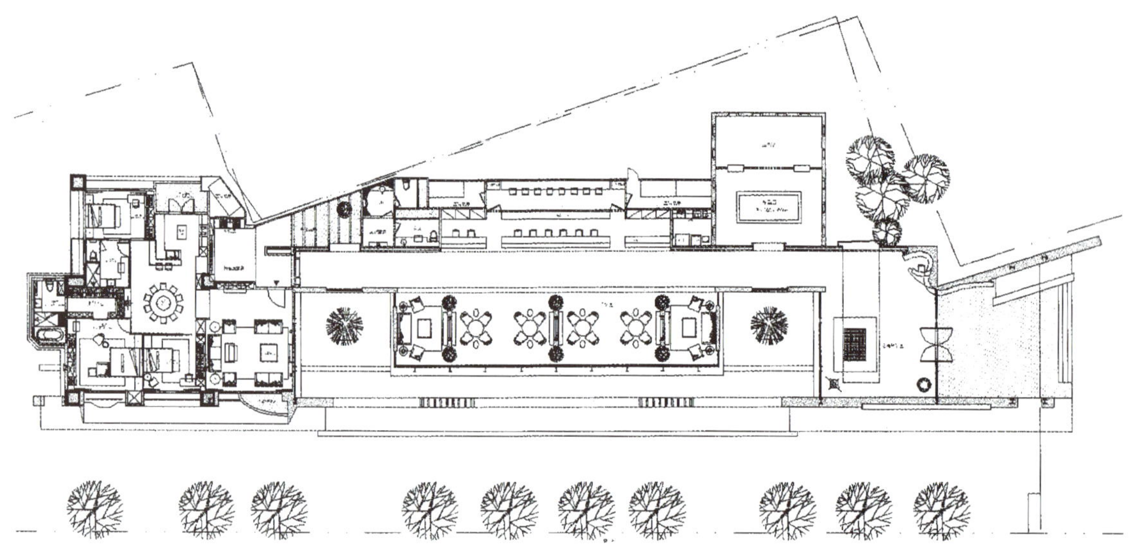

Plan

Left: A kaleidoscope of gorgeous colors of the reception center stands in the green area like a piece of elegant sculpture.
Right: A conrner of the discussion area.

Left: The wall of the corridor is decorated by the tactile stone.
Right: The discussion area is partitioned by oak furnaces, each of which has three metal pipes serving as the exhausts. This kind of design possesses an impression of a social hall.

The reception hall of gallary style displays some giant sculptures, which lowers the business pressure.

Left: House-model area is predominated by black. The glittering metallic lustre shines brightly in the lamplight.
Right: The tactile wall surfaces with graceful beauty sense.

Leather sofas carry their natural nobility. The Thailand teak grilles contrast with the white straight-grained floor.

RIVER RIVER RIVER
BANK BANK BANK
HOME HOME HOME

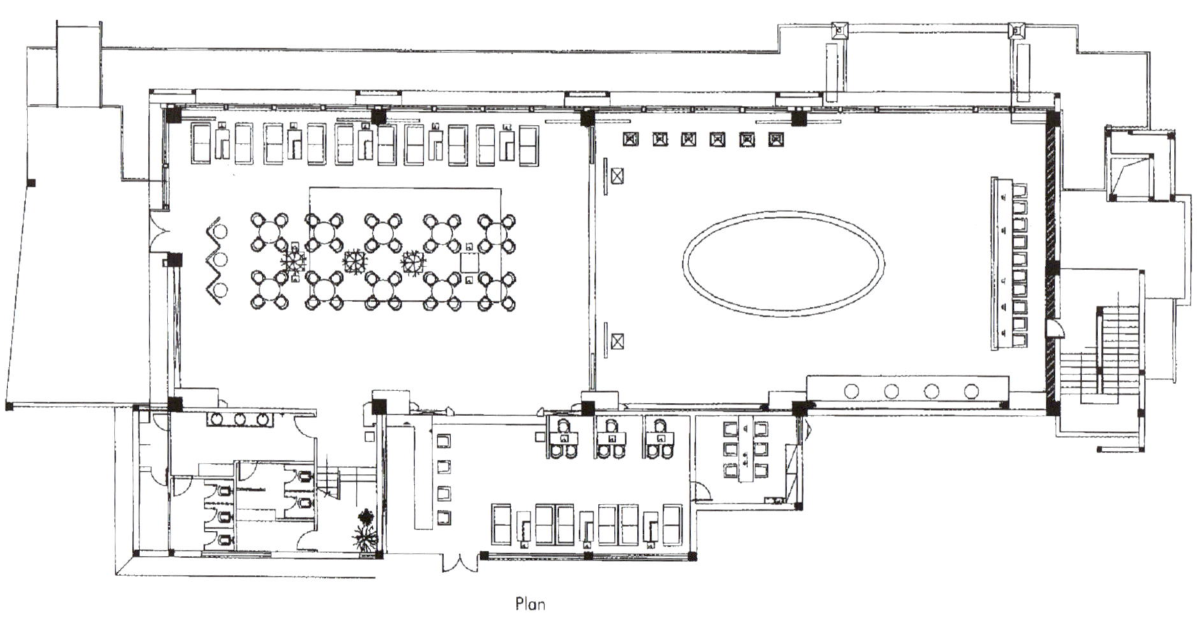

The peaceful figure of the Buddha, splash of sunshine, and pottery with natural gloss are of a special characteristic.

Plan

Rough washbasin with greenish leaves vividly reproduce a Southeast Asian scene.

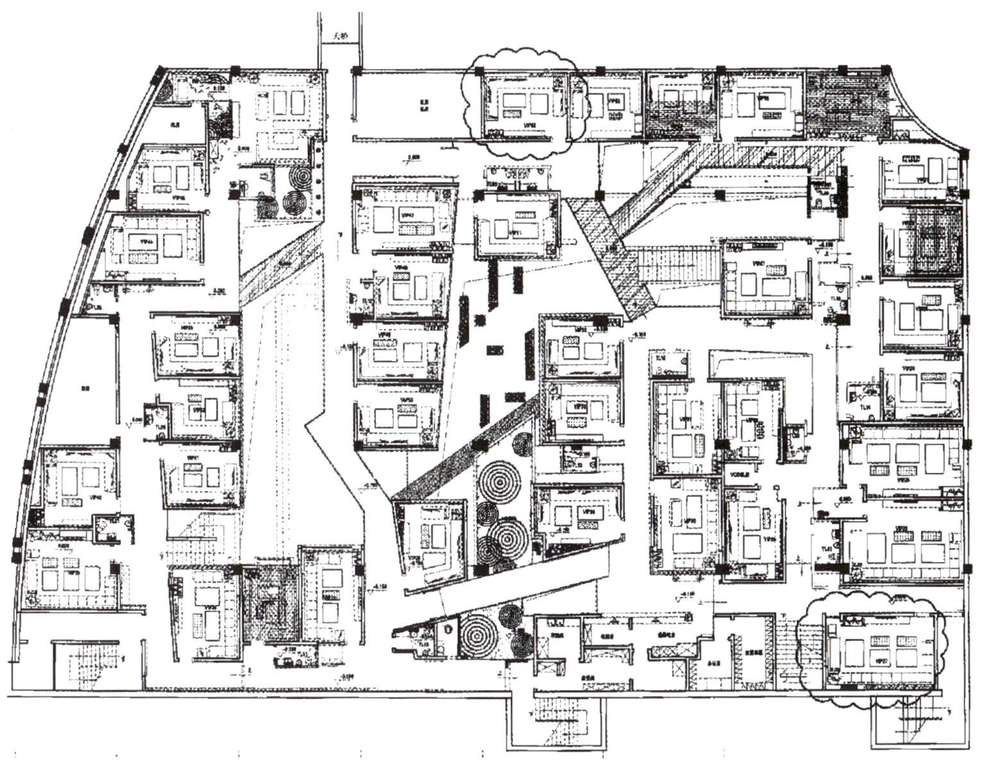

Plan

The terse designing of reception part at the entrance. The marbled floor reflecting the lights from the ceiling.

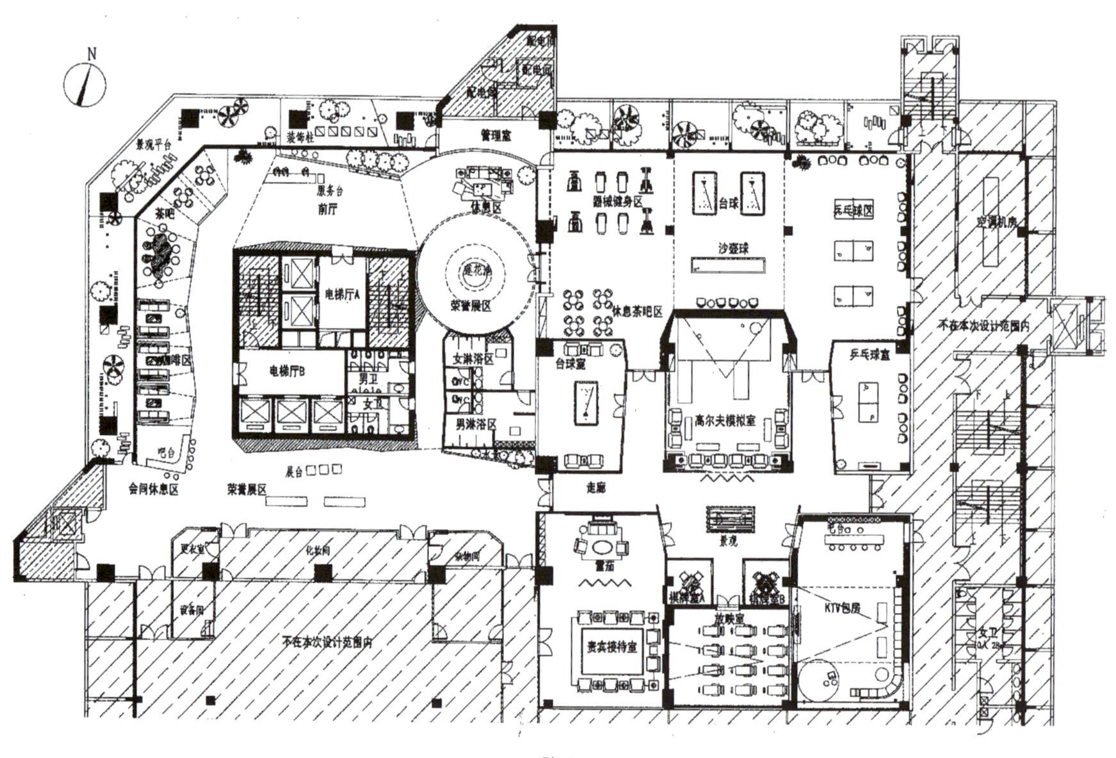

Plan

Left: The front view of the hall form the entrance.
Right: The art decors in the washing room

The washing rooms are finely-fitted
on both sides of the discussion
room, male and female of each
side.

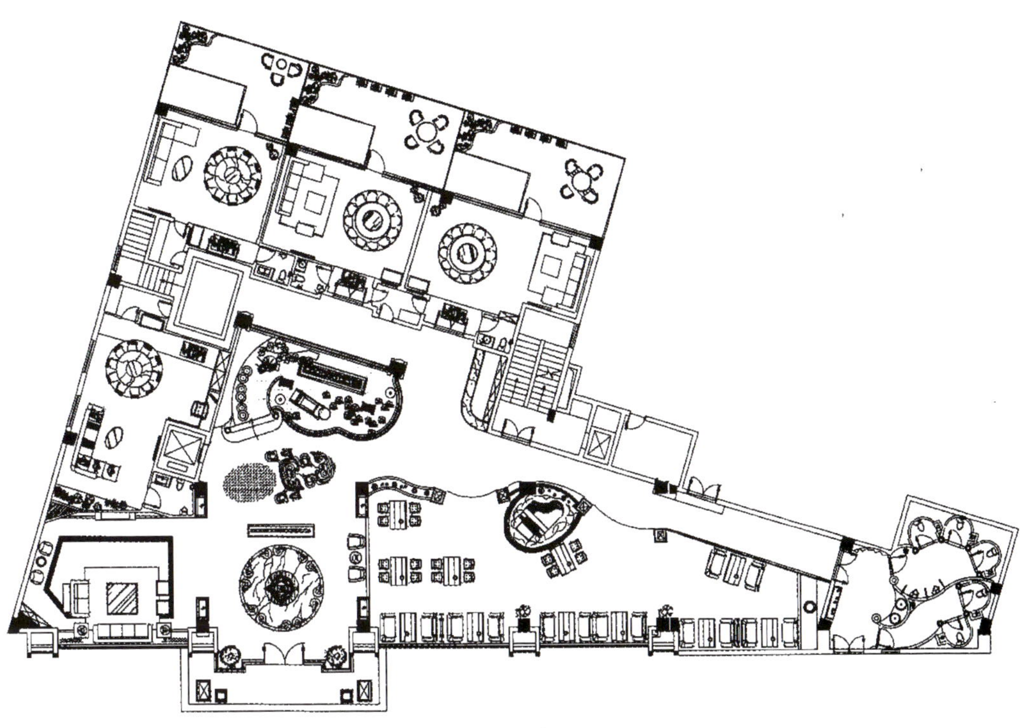

First Floor Plan

Floral lobby from different angle views.

Left: : Flaming pendant ornaments in the atrium is the central axis.
Right: The leisure zone models on a piece of art work.

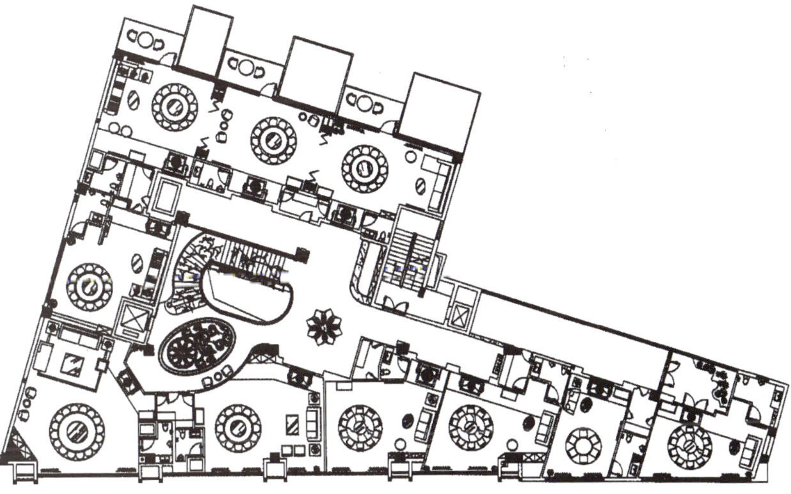

Second Floor Plan

Left: One corner of the hall looks like a mini room showing art works.
Right: Corridor also plays a role in art exhibition.

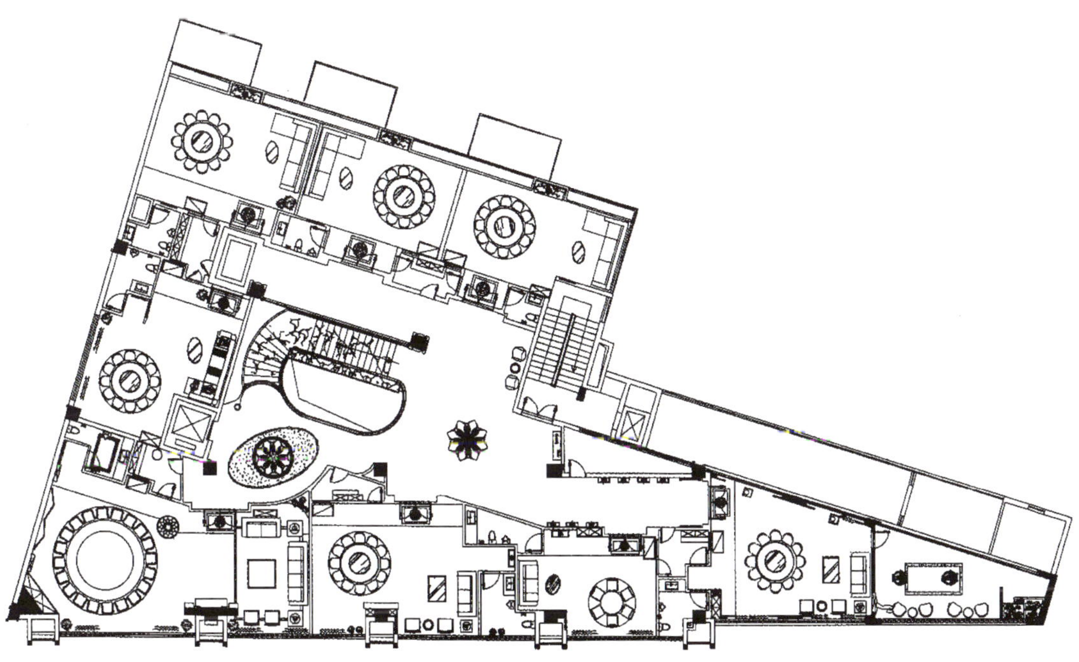

Third Floor Plan

Floral ceilings put on a colorful coat for the lobby.

149

There is a TV in the reserved room, to have a considerate service.

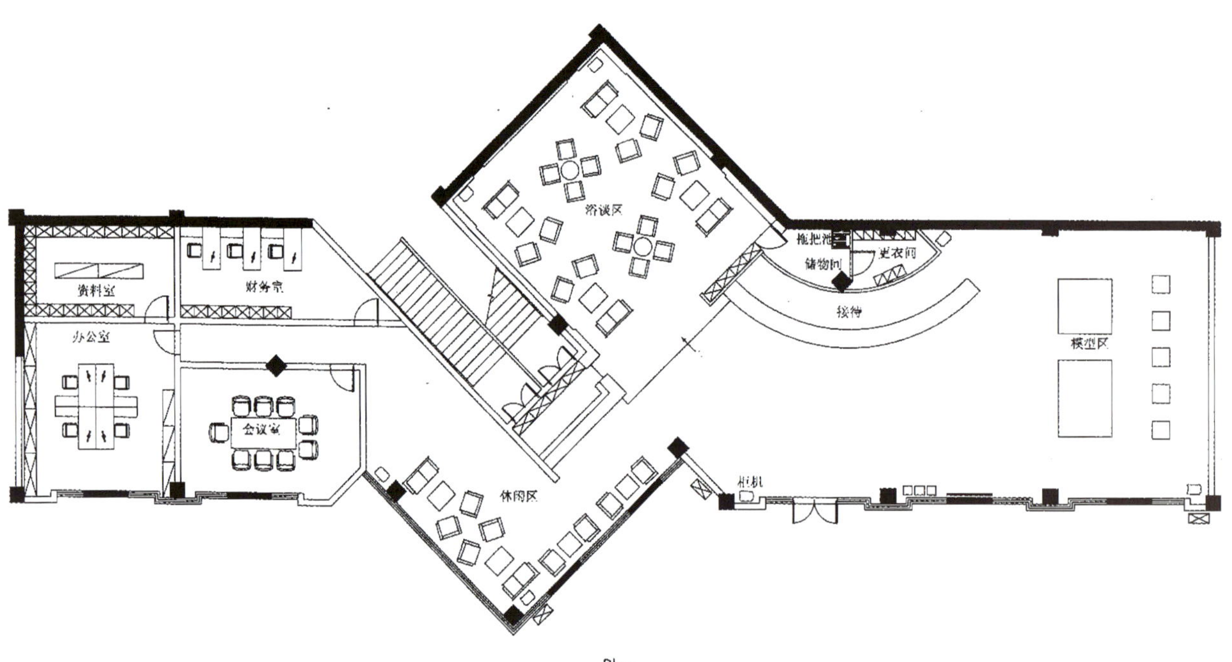

Plan

The exhibition area with mauve gauze curtains blurring the street views.

Fashionable model exhibition area with fine details.

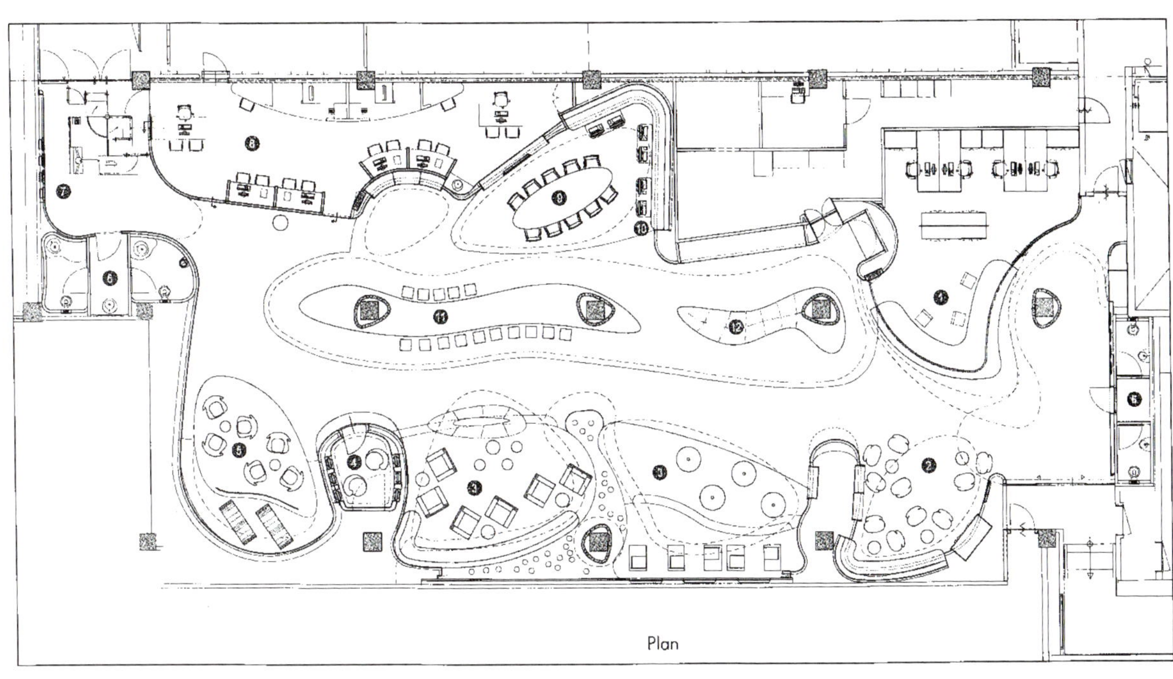

Plan

Left: The club is designed to imitate space capsule.
Right: Rosy pink light adds a tender sense.

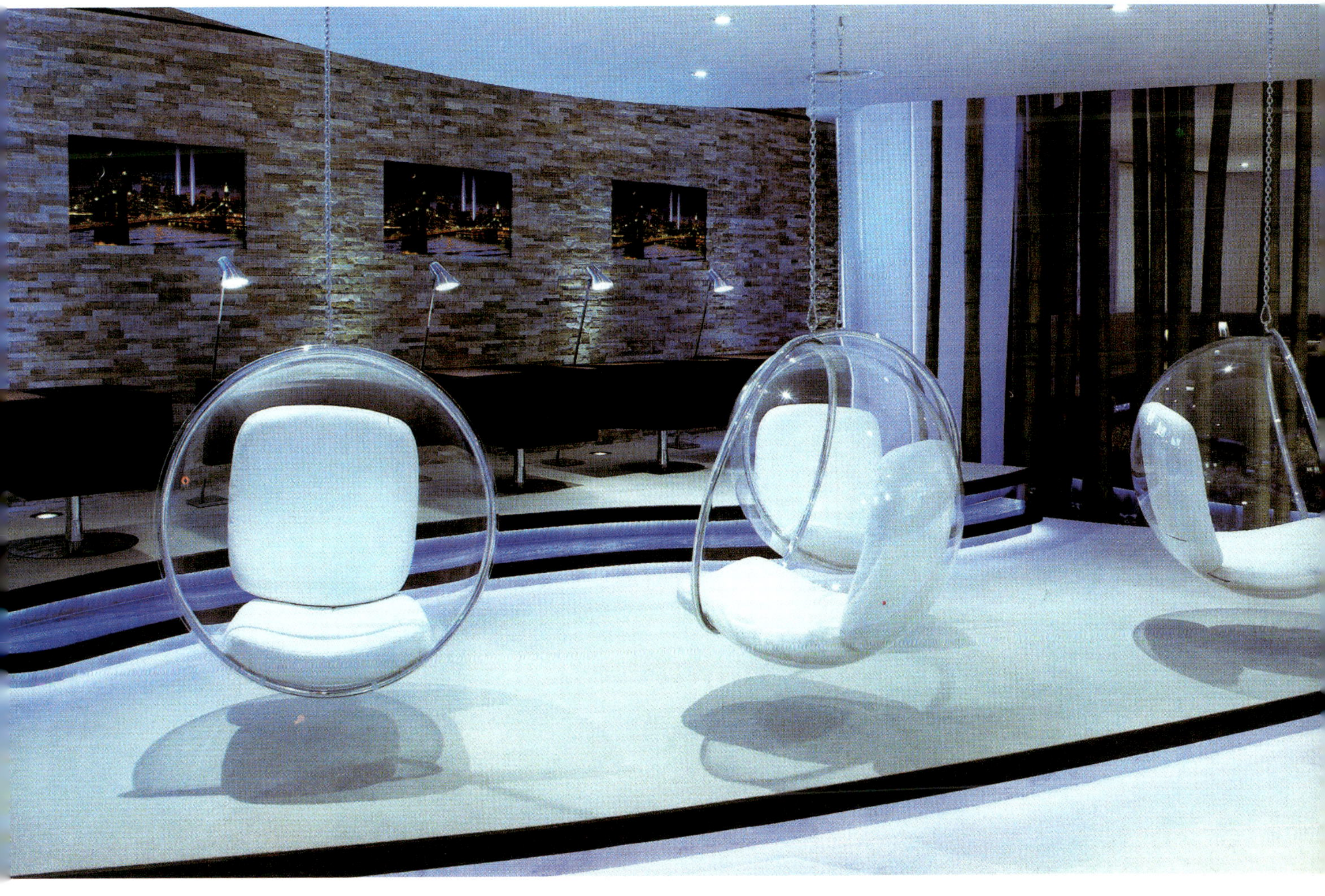

Left: Tables and chairs of tuturistic design.
Right: Chinese bamboos work as partition and set a
      peaceful atmosphere.

Combinations of raw timber and stone materials reproduces a natural scene, and at the same time brings out the Chinese architecture of stone and timber culture.

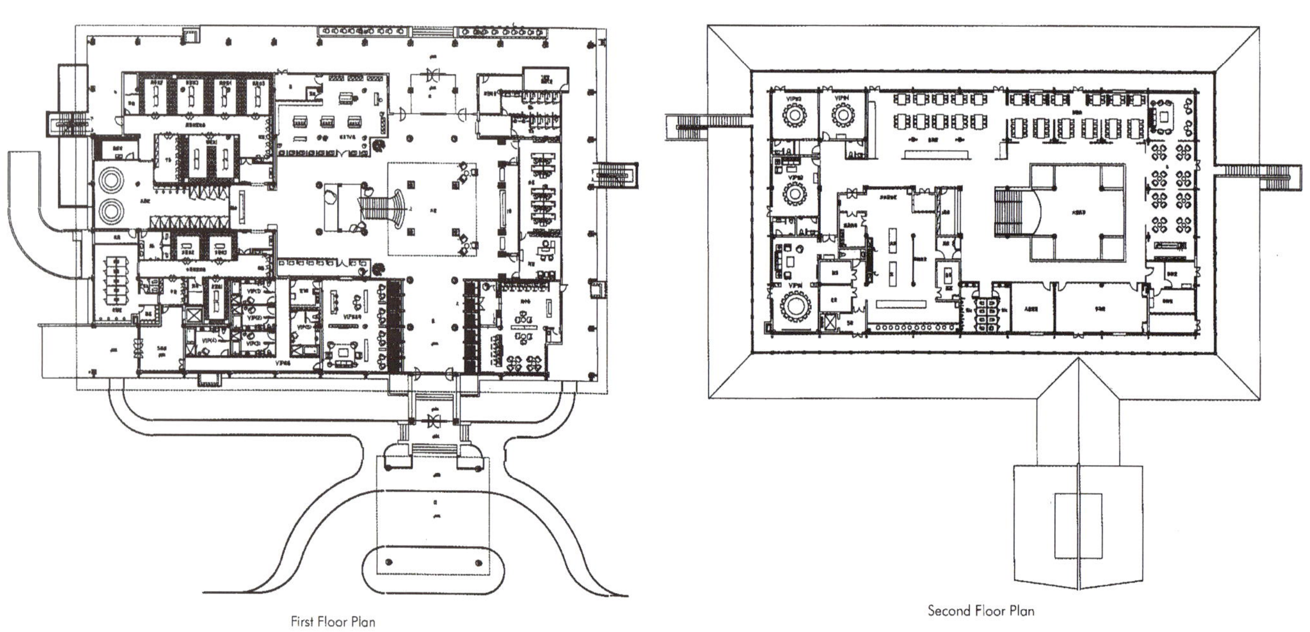

Even the indoor designing has an architectural sense.

First Floor Plan

Second Floor Plan

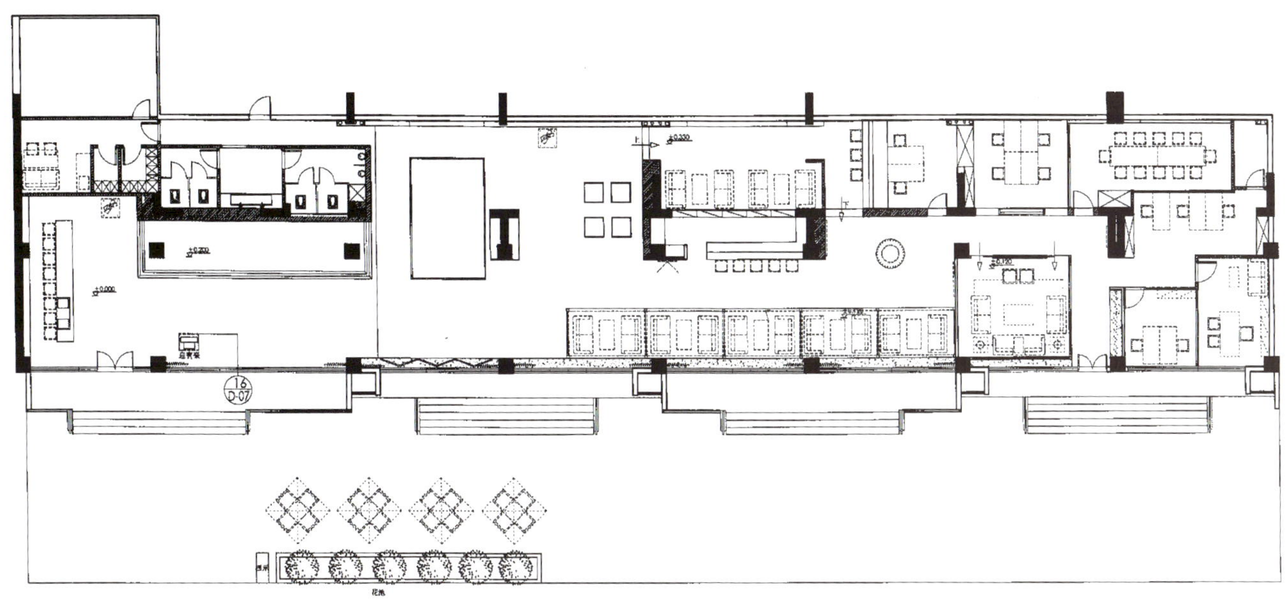

Plan

The two Chinese characters of "Tashan" at the first sight of the entrance, with continous moutain pattern as its background.

Left: Taken a close look at the walls, the stereoscopic
effect fully succeeds.
Right: The indoor design weighs much on the Neo-
Chinese style. Its simple elegance embodies
the humanisticism.

Large paper pendant lamps of traditional Chinese water paintings. Black lampshades make the lights stand out with elegant profoundity igniting the whole peaceful room.

First Floor Plan

Second Floor Plan

Left: The angel painting of the entrance hall highlights the luxurious level of the whole space designing; the floor patterns showcases a masculiar extraordinariness.
Right: The ceiling of the hall is filled with oil-paintings, making a contrast to the golden Buddha at the steps with Oriental favors.

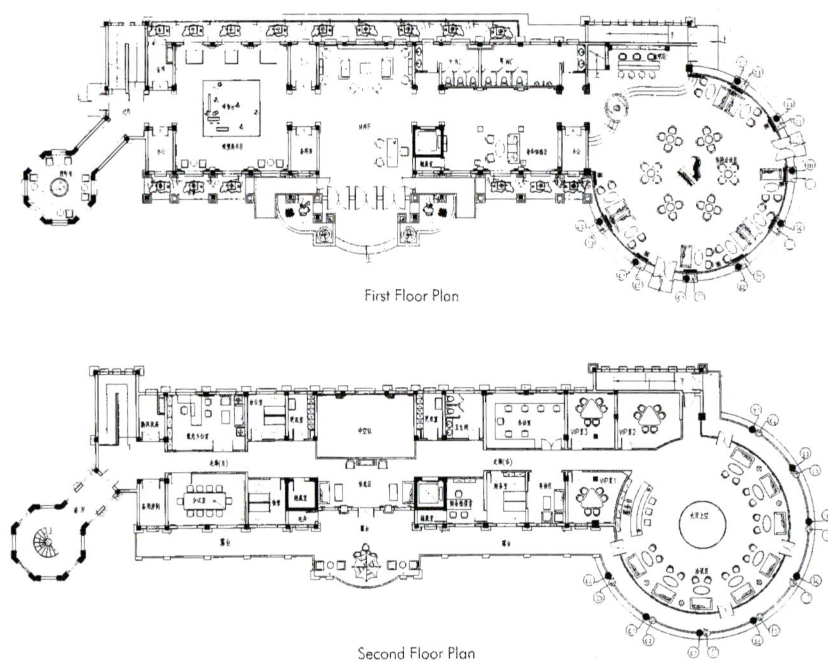

Left: Royal paintings on the walls, precious wines and Western style curtains.
Right: The music coming from the piano under the long pendant lamps makes the space full of romance.

The designing of the reserved rooms
combines Chinese traditions
with European style,
using the classic tones to
show its uniqueness.

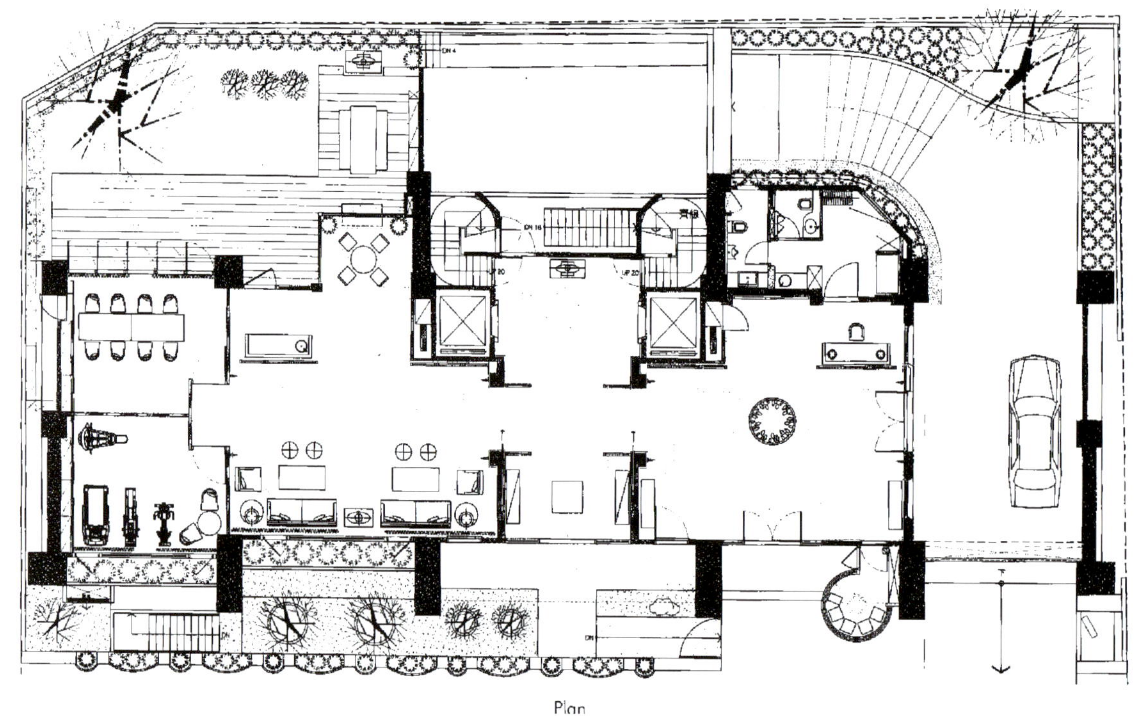

Plan

The front view of the hall,
with classic elegancy and
ordered arrangement.

Left: The passageway with succinct and refined designing.
Right: The round tables in the hall, embelished with floral pots going with the plain background.

Quaint tables and chairs integrated with the quaint floor tiles gives an atmosphere of natural, simple and quiet.

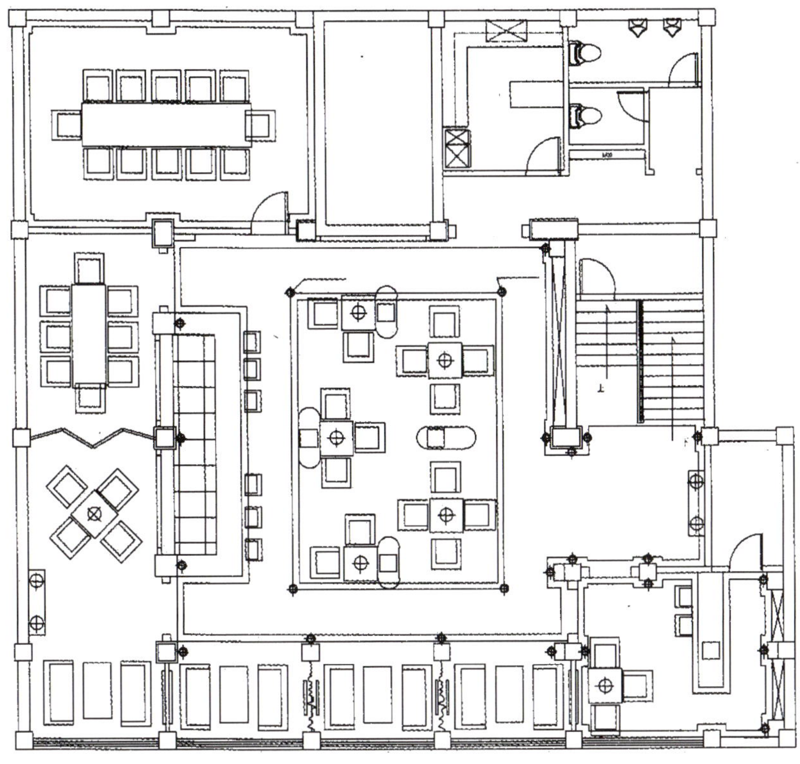

Plan

Left: The reserved box can be used for enterprises' parties and activities.

Right: The reserved box is distinguished but unsophisticated. Having a party with friends here, you may forget any troubles and enjoy your life.